# TELEPHONE
## COMMUNICATION
## SYSTEM ESSENTIALS

# TELEPHONE
## COMMUNICATION
## SYSTEM ESSENTIALS

### TELELIVEZONE

## S. Sudhananthan

PARTRIDGE

A Penguin Random House Company

To order additional copies of this book, contact
Toll Free 800 101 2657 (Singapore)
Toll Free 1 800 81 7340 (Malaysia)
orders.singapore@partridgepublishing.com

www.partridgepublishing.com/singapore

# Preface

This book is titled as Telephone Communication System Essentials. This is first book of Telelivezone series. Telelivezone stands for Telecommunication Living Zone. This book is mainly explaining on how communication runs through the telephone network.

The topics expressed in this book have been carefully oriented with the purpose of providing an integrated picture about the elements involved in telephone communication system. The chapters are organized as a signaling, switching, channeling, communication systems and telephone networks.

Necessary diagrams are presented in certain parts of this book to provide a clear visual understanding of the topics discussed.

The reader can be students, lecturers, professionals or anyone with curiosity to know-how telephone network elements work, enable to understand the key information presented by author.

This book can also be used as a handy manual by professionals working in the telecom fields, where contents in this book can help to strengthen their understanding.

Gathering data for the completion of this book has been very challenging. The author has taken great care in choosing each bit of information to avoid inaccuracy. Additional efforts are also being made in the cross- reference information.

## About Author

S.Sudhananthan is a Telecommunication Engineer with 5 years of experiences in Telco industries relating in mobile communication, landline communication and VoIP communication. He has completed studies in HND Telecommunication (UK) in year 2010. He has keen interest in writing technical books and doing research on innovation and the latest technology.

## Acknowledgement

As author of this book I would like to thank my publisher who was instrumental in making success of this book by designing, formatting and distributing to vendors.

I would like to tell my appreciation to Edit Write Services, who assist to make proof reading and quality checking for this book.

I also would like to express my gratitude to my family and friends for their undivided support to complete this book.

# Table of Contents

# Table of Diagrams

# Introduction

# Telecommunication and the Telephone

Telecommunication has come a long way since the discovery of new breakthroughs in technology. Defined as the transmission of messages and information over long distances through the use of electrical and electronic devices, this process of exchanging information has played a significant role in the enhancement of the world's economy. With telecommunication networks, miscommunications or lack of communication between people in faraway places has now become an obsolete possibility. As long as people have access to telecommunications networks with excellent signals, communication is always just a click away.

One of the most popular types of telecommunication device is the telephone. This communication device allows two or more parties to converse even when they are miles away from each other. The telephone is designed in such a way that it allows the conversion of sounds, such as the human voice, into electronic signals, which are then transmitted in audible forms via wires and cables toward the end users.

The telephone is inarguably one of the most amazing devices ever created. The introduction of this device into the market afforded businesses the opportunity to reach out to more prospects and to better leverage their influence over their target markets. It also impacted the government, households, and individuals on a positive note.

Since the invention of the first telephone by Alexander Graham Bell in the 1800s, new innovations in this technology have sprung up for the world to enjoy. From the conventional telephone patented by Bell in 1876, we now have different types of telephone systems, each designed to meet the needs of different users. While there may be several models of telephones nowadays, though, the main function of a telephone has never changed. The main parts that make up this device remain the same as well despite years of innovation and new breakthroughs in this technology.

How do telephones work and how are messages transmitted through these devices? The fundamentals of these mechanisms are discussed comprehensively in this book.

# SIGNALLING

# Chapter 1

# Signals

## 1.1    Signal Wave Elements

A signal is the main element responsible for conveying information in distance communication. Whether it is landline communication, mobile communication, satellite communication or any communication in between transmitter and receiver its need signal to communicate. It is the energy that causes a circuit to perform its intended action, be it transmission or reception of messages and information.

Signal waves exits in different shapes and lengths. They can be sinusoidal or sine, triangular, square, or saw tooth waves. These different shapes of signal wave used based on their need in carrying or representing information based on their application design.

A signal wave comprises of several elements that consistently vary based on its requirement of information transmission. These elements control the efficiency and intensity of the information being transferred. Signal wave elements include wavelength, period, frequency, and amplitude. These elements are shown in Diagrams 1.1(i) and 1.1(ii), respectively using a sine wave.

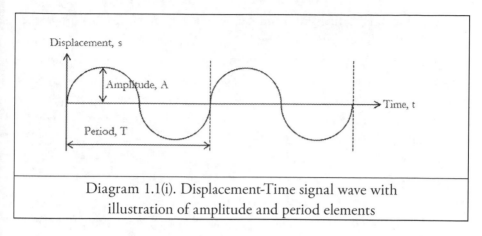

Diagram 1.1(i). Displacement-Time signal wave with
illustration of amplitude and period elements

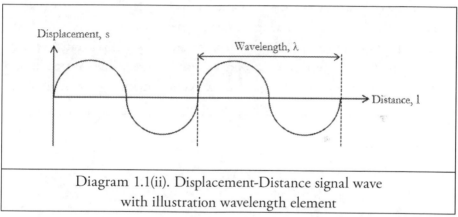

Diagram 1.1(ii). Displacement-Distance signal wave
with illustration wavelength element

The elements shown in Diagrams 1.1(i) and 1.1(ii) are defined as below:

i.   Amplitude -The maximum displacement of a wave from its equilibrium position.
ii.  Period -The time takes to make one complete wave cycle.
iii. Frequency -The number of wave cycles made in one second.
iv.  Wavelength -The distance between two consecutive points on a wave that propagates at the same phase.

Communication signals are not only existed after the advent of electronic communication and the invention of various types of telecommunication devices. In fact, the use of signals can be traced back since the long distance communication among people began. Signals have always been a crucial part of communication, particularly communication over long distances. In the past, signals come in the form of smoke, sound, and drawings. Sending notes using birds as well as lighting up fires at night were also widely used. In today's modern society, these manners of sending signals are long gone, replaced by more modern methods. With the advent of new breakthroughs in technology that started with the invention of the first telephone in the 1800s up to the time of contemporary telecommunication systems in this new era, signals are now classified as analog or digital.

## 1.2   Analog Signal

Analog signal is a type of continuous signal wave with time-varying quantities. The changes of its physical properties and the different values in its waveform have a smooth transition, producing sounds that have no breaks or interruptions. Communication information in the form of analog signal is transferred by means of analog communication channels, such as wire, cable line media and radio frequency waves. Diagram 1.2 shows a coherent analog signal graph.

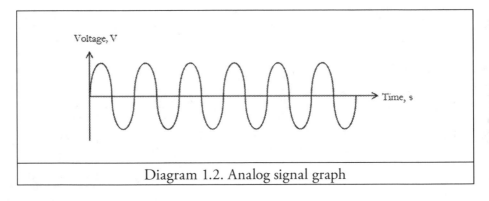

Diagram 1.2. Analog signal graph

Analog signal is commonly used in a variety of applications, the most common are which include the following:

i.   Frequency Modulation (FM) radio broadcasting
ii.  Aerial television broadcasting
iii. Walkie-talkie or handheld transceiver communication
iv.  Amateur radio or ham radio
v.   Sound Navigation and Ranging (Sonar)
vi.  Radio Detection and Ranging (Radar)

## 1.3   Digital Signal

Digital signal is a type of signal that is transmitted in bits pattern. It consists of pulse trains that rapidly change between two levels of intensity. The digital signal can be one of two values, such as 1V for maximum levels and 0V for minimum levels. Unlike analog signal, digital signal has stepping appearance. It may also be square and discrete. The discrete appearance is due to the signal trying to approximate values. At a distance, the stepping appearance may not be readily perceptible and the pulse variations of the signal wave may look analog and smooth. Up close, though, the discrete steps of the signal may become visible. Diagram 1.3 shows a digital signal graph with continuous pulse variations.

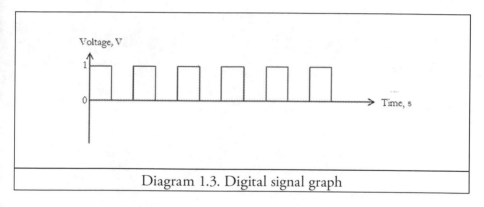

Diagram 1.3. Digital signal graph

Digital signal is also used in a number of varied applications. These include Auto Tele Machine (ATM), Global Positioning System (GPS), Electronic mail, and E-Commerce.

## 1.4   Comparison Between Analog and Digital Signals

Both analog signal and digital signal transmission are varied based on their own characteristic as summarized below:

i.   Analog Signal

    a.   Appears as a continuous wave form with variable amplitude.
    b.   Vulnerable to noise interference during the transmission process.

    c. Restoration of the original signal transmitted is not feasible.

    d. Information is presented in sine wave form.

ii. Digital Signal

    a. Information is represented as pulse train and transmitted in discrete values as 0 and 1.

    b. Can be immune to noise during the transmission process.

    c. Allows easy restoration of original signal.

# Chapter 2

# Signaling

## 2.1 Overview of Signaling

Signaling is the process of sending signal throughout networks. This process is an important aspect in telecommunication and telephone industry in general. Without signaling it would be difficult to transfer information from one point to another. In a nutshell, signaling can be said as nerve of any network mechanism.

Signaling is progression of a moving element from one place to another through wired or wireless communications channel. In a standard telephone network, signaling is the process of exchanging information for the purpose of establishing, maintaining, and terminating telephone call connections. A standard signaling block diagram is shown in Diagram 2.1.

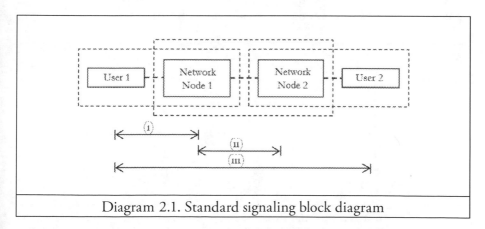

Diagram 2.1. Standard signaling block diagram

8

Three categories of signaling are used in telephone networks. These are the following:

i.   User to network node signaling.
  a.   A type of signaling used to ensure the standby, establishment, and active status of telephone calls.
ii.  Inter-network node signaling
  a.   A type of network signaling used for establishing and controlling the exchange of information among various networks.
  b.   It is used to oversee time allocation, supervise billing, and improve organizational management.
iii. User to user signaling
  a.   A form of signaling employed for special operations, such as in Dual Tone Multi Frequency Signaling (DTMF) and Unstructured Supplementary Service Data (USSD) code.

## 2.2   In-Channel Signaling

In public switched telephone network (PSTN), in-channel signaling is the process of sending signaling information and voice information within the same channel. Known also as in-band signaling, this method encodes and transmits telephone numbers as Dual-Tone Multi-Frequency (DTMF) tones. With the help of the DTMF tones, this process also provides inter-exchange telephone companies instructions on how to route the calls. Below is a visual model of in-channel signaling, where both voice and signaling information are transmitted or sent simultaneously within the same channel.

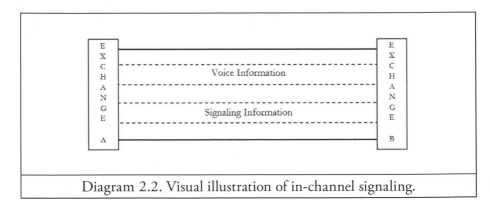

Diagram 2.2. Visual illustration of in-channel signaling.

Since the path of voice information and signaling information are the same, certain problems may occur. For instance, signaling may interfere with voice data during active voice calls or it may clog the path of the voice information being sent. This occurrence can result to congested signals, busy tones, or incomplete calls. Sharing of the path by voice and signaling information can ultimately slow down the call setup.

## 2.3   Common Channel Signaling

Common channel signaling (CCS) is a transmission method that follows a process different from that of in-channel signaling. In this type of signaling voice information and signaling information each will have a dedicated and separate channel for encoding and transmitting. A visual illustration of common channel signaling is shown in Diagram 2.3.

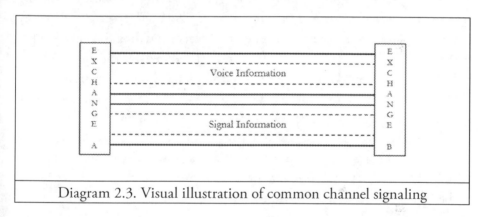

Diagram 2.3. Visual illustration of common channel signaling

Unlike in-channel signaling where the call set-up may be slow because of the clogging of paths, the transfer of information in CCS is faster and simpler. There is also no interference of voice signals in this type of signaling. Furthermore, the call setup and switching processes are done effectively by maintaining the privacy of information throughout the transmission process.

# Chapter 3

# Signaling System Number 7 (SS7)

## 3.1    Overview of Signaling System Number 7 (SS7)

Signaling System Number 7 (SS7) is a type of common channel signaling system that was first developed and used in 1983. This set of telephony signaling protocol was primarily created to replace Signaling System Number 6 (SS6), the first international CCS protocol defined by the International Telecommunication Union Telecommunication Standardization Sector (ITU-T). Both the SS6 and SS7 were called Common Channel Signaling systems or Common Channel Interoffice Signaling systems (CCIS) because of their ability to efficiently separate signaling and bearing channels. Signaling System 6 though, was not responsive to digital systems and had limited function with its 28-bit signal unit. The creation of SS7 improved the speed of signaling and enhanced the holding time of the bearer channels.

The primary use of SS7 is to synchronize call setups, call routing and call management processes in a telephone network. Unlike its predecessors, it uses the outbound signaling method, which uses a dedicated and separate channel for signaling and voice transmission during voice calls. Because of the presence of dedicated channels, outbound signaling makes routing call and remote network management more efficient. Due to its efficiency and reliability, SS7 is now being used as the primary mechanism for new installations worldwide. It is utilized as the interoffice signaling protocol for Integrated Services Digital Network (ISDN) and for other standards outside of ISDN. A visual of model of the SS7 network is shown in Diagram 3.1.

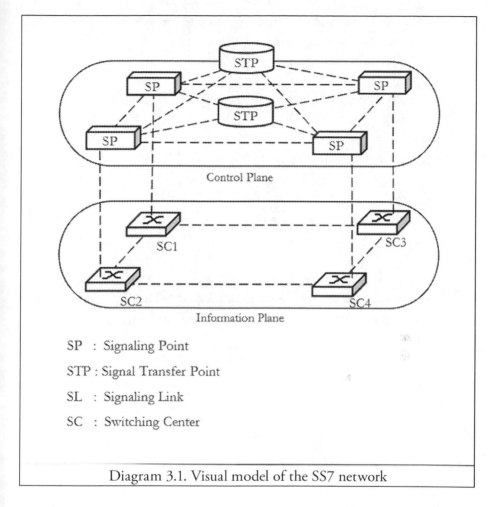

Diagram 3.1. Visual model of the SS7 network

Signaling System No. 7 is made up of two essential planes. These are the control plane and the information plane. These planes serve different functions, which are vital to the mechanism of SS7. The functions of each plane are explained below:

i. Control Plane
    a. Control plane is responsible for the establishment, preservation, and management of call connections as requested by the user.

ii. Information Plane
    a. Information plane is specifically designed for the exchange of information between users via multiple switching centers.

The control plane is comprised of three important elements, each of which has dedicated functions. These elements include:

i.   Signaling Point
   a.   Signaling point is the network node that directly links to the switching center.
ii.  Signal Transfer Point
   a.   Signal transfer point is responsible for routing the incoming signaling information to the necessary destination.
iii. Signaling Link
   a.   Signal link is the link that interconnects each of the other signaling points in SS7.

Since the development of the SS7 technology, many advantages have been experienced in the field of telephone communication. Signal System Number 7 has made a big transformation in the quality and productivity of call setups.

# Chapter 4

# Bandwidth and Data Rate

## 4.1 Bandwidth

Bandwidth refers to the capacity of a transmission channel to carry information and transfer it from one network to another. It is the measure of the maximum amount of communication data can be sent from an individual channel within a particular moment. In the field of telephony signaling, particularly in analog communication systems, bandwidth is measured in Hertz (Hz). A visual illustration of bandwidth in an analog transmission channel is shown in Diagram 4.1 (i).

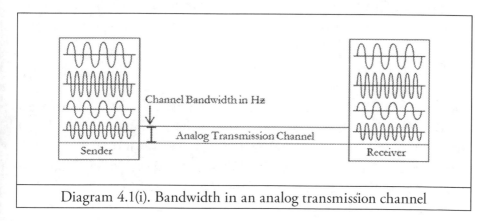

Diagram 4.1(i). Bandwidth in an analog transmission channel

In the digital communication system, bandwidth is being measured in bits per second (bps). A visual representation of bandwidth in a digital transmission channel is shown in Diagram 4.2 (ii).

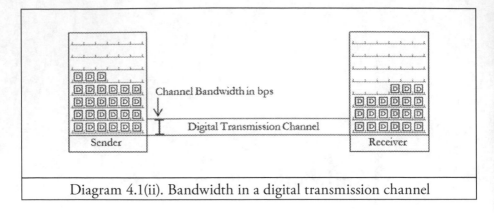

Diagram 4.1(ii). Bandwidth in a digital transmission channel

When excessive information is being sent, the bandwidth may become overloaded. This will limit the rate of information being transferred and lead to the delay in the transmission of information. To understand this better, let us take the diameter of a water pipe as an example.

Let us assume that the diameter of the water pipe is the bandwidth and the volume of water that goes through is the information being transferred. The bigger the size of the pipe, the more water will be able to flow through it within a given moment. The smaller the pipe, the lesser amount of water will be able to flow through it. The bandwidth in telecommunication is much like the pipe. The bigger the bandwidth, the more information will be able to go through it. The lesser the bandwidth, the lesser amount of information will be transmitted.

## 4.2   Data Rate

Data rate refers to the speed at which communication data is transferred between telecommunication networks. The volume of digital data transferred or handled by a particular network is measured by the quantity of bits that pass through it. Data rate is measured in bits per second (bps). A visual illustration of data rate in a digital transmission channel can be seen in Diagram 4.2.

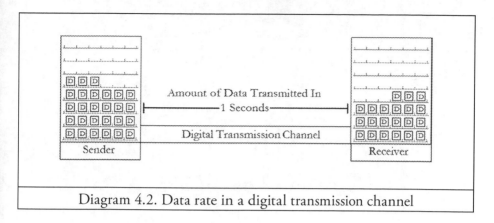

Diagram 4.2. Data rate in a digital transmission channel

# Chapter 5

# Gain and Loss of Signal

## 5.1    Gain and Loss of Signal

In telephony signaling, gain of signal refers to the increase in signal strength during a particular transmission process. Known also as signal amplification, it pertains to a signal's increase in power or amplitude as it is being transferred from the source to the intended destination. Signal gain is measured in logarithmic Decibel (dB) units. Diagram 5.1[i] shows visual representation of a normal signal and an amplified signal, respectively.

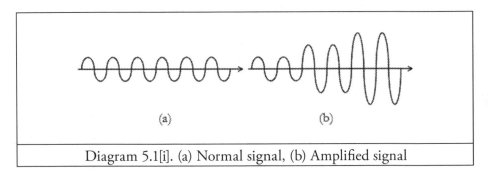

Diagram 5.1[i]. (a) Normal signal, (b) Amplified signal

Loss of signal (LOS), on the other hand, means the reduction of signal strength during a particular transmission process. Known also as signal attenuation, it can be stated that the connection between telephone networks is gradually losing. Loss of signal can be due to a variety of reasons, including range problems, improper network configurations, and interference to name a few. A visual representation of a normal signal and an attenuated signal is shown in Diagram 5.1[ii]

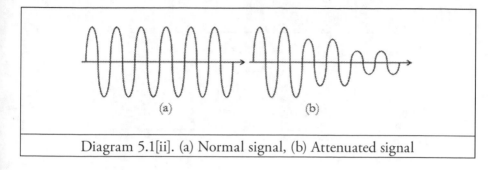

Diagram 5.1[ii]. (a) Normal signal, (b) Attenuated signal

## 5.2   Signal Amplifier

A signal amplifier or signal amp is basically a device that is used to boost the strength of an analog signal, keeping it from attenuating during a particular transmission process. It is also known as a signal booster. Diagram 5.2 shows a visual representation of an analog signal amplification process using a signal amplifier.

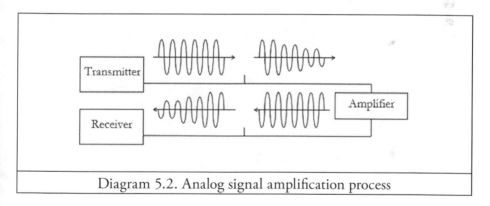

Diagram 5.2. Analog signal amplification process

## 5.3   Signal Repeater

A signal repeater is a device used to regenerative signals at a relatively higher power or higher signal than their original forms. A signal repeater is used to recover digital signal from impairments by regenerating the digital signal and retransmit to reach its extended destinations. A visual representation of a digital signal being regenerated using a signal repeater is shown in Diagram 5.3.

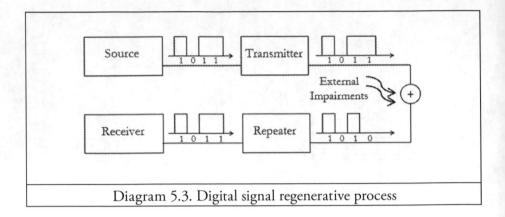

Diagram 5.3. Digital signal regenerative process

# SWITCHING

# Chapter 6

# Switching

## 6.1    Overview of Switching

The world's first telephones came in pairs and were often used only for private use. Unlike today, where people can make calls to virtually anybody around the world, the first telephone prototypes only allowed calling between two paired networks. If people needed to talk to other entities, they would have to buy another set of pairs to make contact.

In 1889, calling to other phone numbers apart from the paired units was made possible through the invention of the switch by Almon Brown Strowger. Called "The Strowger Switch", this device allowed calls to be routed to other telephone lines. The switchboard became a hit and human telephone operators then became a crucial element in routing calls during this period.

The first switch invented by Strowger became the prototype for the telephone switches that are now being used in today's modern telephone. Now switches, which are basically interlinked nodes, have become a fundamental and vital part in the present lines of communication networks. Unlike the early models, though, which made use of human operators to route calls, today's switches are capable of automatically routing calls to intended destinations.

A switch is basically a device that can route signals from one telephone line to another. It plays a crucial role in telephone exchange, interconnecting subscriber lines and establishing telephone calls between subscribers regardless of distance. The process of routing signal information from one switch to another is called switching.

The switching process is better presented in the basic visual diagram illustrated in Diagram 6.1. Here, multiple paths made up of different selective nodes for exchanging information between X and Y is shown.

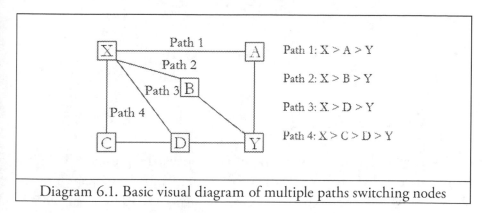

Diagram 6.1. Basic visual diagram of multiple paths switching nodes

As seen in Diagram 6.1, information can be sent through any of the paths that exist. However not all paths are have the same level of efficiency.

Several factors are taken into consideration in the selection of switching paths. These factors include capacity, delay, error, and cost, which are explained below:

i.  Capacity- The capacity of the path to handle information at the time it needs to establish a connection.
ii.  Delay- The amount of time it takes for information to reach its destination through the selected path.
iii.  Error- The amount of errors that information may encounter while passing through the path.
iv.  Cost- The total estimated charge incurred for using a respective path.

Signal switching in communication system can be categories into two. They are circuit switching and packet switching.

# Chapter 7

# Circuit Switching

## 7.1    Circuit Switching

Circuit switching is the most common type of switching used in building a communications network. This switching method used to establish direct physical connections between the communication parties by linking to multiple network nodes. The selection of network nodes will base on the intended destination of information transmission and availability of network nodes. Circuit switching is used in the telephone industry for ordinary telephone calls as only analog signal can be transmitted through this switching process. A visual illustration of circuit switching network is shown in Diagram 7.1.

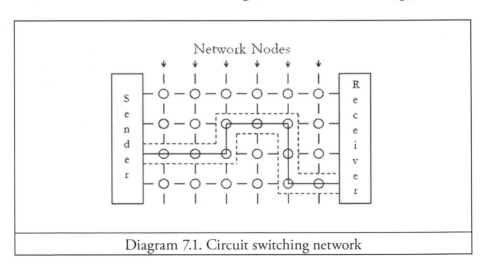

Diagram 7.1. Circuit switching network

The bandwidth of the physical channel path determines the size of the analog data that can pass through a particular switching network. Most of the common applications that use circuit switching include PSTN calls, local area network (LAN), cable television broadcasting, and residential security system.

# Chapter 8

# Packet Switching

## 8.1  Packet Switching

Packet switching is a switching method specifically used for transmission of digital data across the data networks. In this process, data is broken down into blocks that are suitably sized for easier, faster, and more efficient transfer. The blocks are called packets, and they can potentially vary in length, depending on need. The packets are routed by various network mechanisms toward their target destinations.

Each packet contains information such as the sender's address, receiver's address, and sequence number. These bits of information are used by the system to determine the best way to transmit packets to their intended destinations.

Packet switching is divided into two classifications. These are virtual packet switching and datagram switching.

## 8.2  Virtual Packet Switching

Virtual packet switching, known also as network oriented packet switching, is a method wherein packets are routed to certain paths during a communication process. For routing to take place, the network first pre-plans the routing of the packets, considering certain factors, such as the availability of the path, condition of network, and network congestion. Once a path is found, it will then assign this path as the route for packets transmission. Paths used in virtual packet switching are called virtual path. In virtual packet switching the connection may perceived as dedicated physical circuits, but certain parts of the path may also be used simultaneously by other communication systems.

Virtual packet switching uses numbers instead of full destination addresses. This makes this method advantageous as the use of numbers requires less bandwidth and can relatively free up enough space in the available paths. This method is also more cost-efficient. In addition, since this method of packet switching is connection orientated, it is also more reliable in terms of connection. Below is an illustration of a virtual packet switching network.

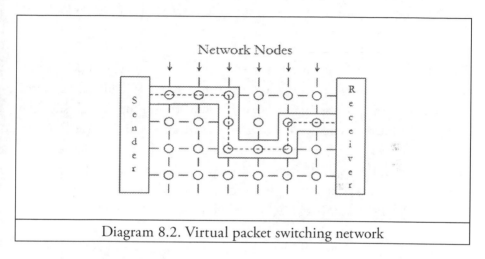

Diagram 8.2. Virtual packet switching network

During the transmission process, all packets are transmitted from the source to the destination using the same route. Packets are transmitted in an orderly manner. Once the connection has been established and the transmission process is completed, the connection nodes or virtual circuit is then cleared and prepped for use for the next packet transmission. This does not mean, though, that the same path may be used when a sender decides to send another message to a receiver. The next messages sent by the same sender may use a different path through an entirely different circuit even though the circuit may have been instantly cleared after the sender's first message went through.

Virtual packet switching is used in a wide variety of applications, including Integrated Services Digital Network (ISDN) calls, local area network, satellite broadcasting, and mobile communication network.

## 8.3   Datagram Switching

Datagram switching, known also as packet oriented switching, is a packet switching mechanism that treats packets as separate entities. Basically, the packets, known as datagrams in this technology, are made to travel independently through the network. Unlike the virtual packet switching methodology, this technology does not use numbers and, instead, rely on full destination addresses to successfully transmit information data. In addition, datagram switching does not rely on pre-planned routes; rather, it selects network paths based on the condition and congestion of networks in all instances.

Each of the datagram in this method is directed to the intended destination of the message through the programmed full destination address. Since the data stream of the packets may follow different paths during the transmission process, the message may not arrive at the destination in proper order. Subsequently, reassembly of the packets has to be performed at the receiver's end to make out the original intended message being sent. A visual presentation of datagram transmission at first time sequence and second time sequence are shown in Diagrams 8.3 (i) and (ii), respectively.

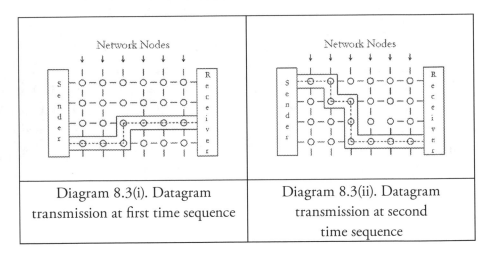

| Diagram 8.3(i). Datagram transmission at first time sequence | Diagram 8.3(ii). Datagram transmission at second time sequence |

## 8.4   Comparison between Circuit Switching and Packet Switching

The characteristics of circuit switching and packet switching make these types of switching distinctly different from each other. Below are their differences:

i.   Circuit Switching

    a.   Used for analog data transmission.
    b.   Bandwidth is pre-located for data transmission.
    c.   Circuit capacity is not susceptible to other network traffic.
    d.   Cost for data transmission is relatively high.

ii.   Packet Switching

    a.   Used for digital data transmission.
    b.   Bandwidth is dynamically allocated based on demand.
    c.   Susceptible to other traffic in the same network, leading to a delay in transmission.
    d.   Cost for data transmission is low.
    e.   Packets are capable for recovery of erroneous data by re-transmission process.

# Chapter 9

# Telephone Switching Method

## 9.1    Stored Program Control

Stored Program Control (SPC) is a technology in telecommunications that is widely used in telephone exchange systems. This microprocessor is the driving mechanism behind electronic switching systems (ESS). With this technology, the routing of calls, establishment of connections, and any other functions of the switching systems in telephone exchanges are stored and control by computerize systems.

Created in 1954 by Erna Schneider Hoover for Bell Laboratories, SPC enhanced the reliability and versatility of electronic switching systems, even enabling sophisticated calling features that made the telephone communications industry an even more beneficial technology to the world. The introduction of the stored program control has paved the way for the integration of all process control as a telephone exchange. It also created a centralized system for all customer-related information. With a centralized system, the switch can now handle thousands of calls between multiple exchange networks. The introduction of time division multiple access (TDMA) and space division multiple access (SDMA) are increase the capacity of SPC telephone network.

SPC manages the signaling and switching during a call. It also handles the administrative processes involved in making calls and billing management. Overall, it has proven to be a reliable control program. Due to the centralization of processes in SPC, any breakdown issues or interruptions to the connection in calls can easily be located, recovered and fixed.

# CHANELLING

# Chapter 10

# Wired Channel

## 10.1 Wired Channel Overview

In communication system channels are playing key role in conveying or transmitting voice and data information through network. Channels are mainly groups to guided and unguided; in other term known as wired and wireless. Wired channel is the most common type of channel used in telephone systems to transmit voice and control information. A wired transmission channel is a physical pathway that provides direct connection between network nodes. Normally, wired channels have coverings that protect them from imminent damage.

Wired transmission channels are categorized into two types. These are baseband transmission and broadband transmission. The difference between the two lies on their capacity and ability to transmit signals efficiently. Known also as a lowpass channel, a baseband transmission channel can transmit signals in full capacity over a network. It can even transfer extremely low frequencies that are almost at the zero level. For efficient transfer, though, the signal must come only from a single source. A broadband transmission channel, on the other hand, can transmit signals in full capacity over a network from multiple sources simultaneously. It can also transfer signals to multiple directions in one setting. Broadband transmission channels have more capacity compared to baseband transmission channels.

A wired transmission channel has an advantage over wireless channels in terms of the quality of transmitted signals. Fundamentally, it can carry huge amount of data with less interference. This type of channel, though, also has its own drawbacks, which make wireless connections a more feasible choice for some people in the telecommunications industry.

One of the disadvantages of wired channels is that they require greater size of space for installation and setup. They may also need quite a number of cables and towers, depending on distance and geography. In addition, physical maintenance which can be cumbersome, is regularly required for this channel to provide continuous, uninterrupted service.

The most common types of transmission cables used in telephone exchange include the twisted pair cable, coaxial cable, and fiber optical cable.

## 10.2  Twisted Pair

The twisted pair cable is the most common type of cable used in connecting telephone networks for homes or offices. It is made up of two independently insulated copper wires intertwined together in spirals. The spiral form of the intertwined cables is designed to help reduce the occurrence of electromagnetic interference and crosstalk induction in the cables during data transmission. Both wires in the pair are required for the connection. Twisted pair cables come in two types. The first, shield twisted pair (STP), is most often used in business locations. The second, unshielded twisted pair (UTP), is most often used in home installations. A visual model of both types of twisted pair cables is shown in Diagrams 10.2 (i) and (ii), respectively.

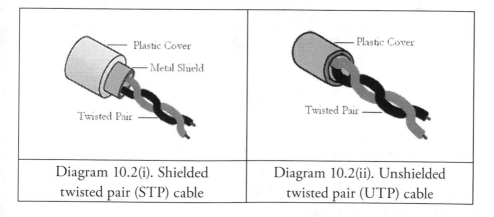

| Diagram 10.2(i). Shielded twisted pair (STP) cable | Diagram 10.2(ii). Unshielded twisted pair (UTP) cable |

As seen on the diagrams, STP and UTP cables have a difference when it comes to the shield that covers them. A shielded pair has an extra metal layer below its plastic cover while an unshielded pair only has the plastic for covering. The metal layer in the STP provides extra protection to the wires as well as the ability to better filter noises that may come from external radiation sources.

Twisted pair cables are classified into seven categories by the American National Standards Institute/Electronic Industries Association (ANSI/EIA). Classification is based on the cables' ability to support bandwidth ranging between 0.4 MHz to 1000MHz. Twisted pair cables under a particular category is used for communication depending on the specific requirement of networks.

## 10.3   Coaxial Cable

A coaxial cable is a type of wire that is thicker than most ordinary wires. This cable consists of a copper wire that is well-insulated by a tubular layer of insulation material. The insulation material is further reinforced with a tubular grounding shield, as well as an outer sheath or insulating jacket. The grounding shield in the cable reduces distortion of data during transmission via the main copper wire. A visual illustration on how the inner layers of a coaxial cable look like is shown in Diagram 10.3.

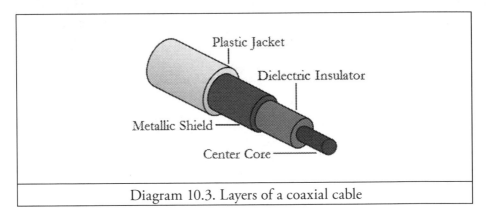

Diagram 10.3. Layers of a coaxial cable

Coaxial cables can support both analog and digital transmissions with up to 1 GHz bandwidth. They are widely used by telephone companies for the transmission of long distance signals. They are also mostly used primarily by local computer networks and by the cable television industry for signal distributions. Amplification of coaxial cables is required for every 10 kilometers when transmitting analog signals.

## 10.4   Fiber Optics

Fiber optics cable is a type of wiring that is made up of glass threads, known as fiber glass, bundled together. The optical fiber glass is thinner than human hair and is capable of transmitting signals, which are modified into light waves. The ability of the cable to use internal reflections of laser lights to transmit data makes it a good option for data transmission, since it allows minimal susceptibility to errors. It also provides good protection against interference during data transmission.

A visual illustration showing the appearance of the inner layers of a fiber optic cable is shown in Diagram 10.4.

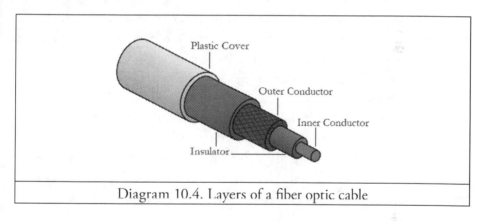

Diagram 10.4. Layers of a fiber optic cable

Apart from its strong ability to protect against interference, fiber optic cables are also known to provide more advantages than any other types of communication lines. One big advantage is that it can carry more data because it has a bigger bandwidth access of around 100Gbps.

Fiber optic cables are rapidly replacing the traditional UTP cables for data and voice transmission in homes and offices. It is utilized widely in interconnecting international telephone networks via the undersea routes. Nonetheless, these types of wire connection also have its disadvantages. It requires specific connectors as well as skilled workers during installation and it also costs more than the other types of wiring connections. They are also more fragile than coaxial cable and twisted pair cable.

# Chapter 11

# Wireless Channel

## 11.1 Wireless Channel Overview

A wireless transmission channel is a channel that can transmit messages over various networks through air or space without the need for a physical path. This type of transmission channel is widely used in telecommunications these days and it has proven to be indispensable for long distance transmissions. Compared to wired transmission channels, wireless transmission channels are relatively more economical. They are also easier to install and maintain. Wireless channels also have their own limitations. Unlike its wired counterpart, this type of connection is prone to the effects of electromagnetic and solar radiation. It is also susceptible to distortion, particularly when passing through an inconsistent geographical area. Furthermore, a wireless channel connection may experience serious problems in signal quality during bad weather.

The most common types of wires transmission channels that right in use in our daily life are infrared, radio wave and microwave.

## 11.2 Infrared

Infrared is a type of wireless communication channel that uses the 'line of sight' (LoS) technology, a propagation that transmits and receives data only when both the sending and receiving mechanisms are in line of site of each other, for transferring data. Limited by geographical obstructions, this type of transmission channel is used only for short distance communication. For this communication channel to work, the sending and receiving stations must only be a few meters away from each other. Infrared technology has been used widely to power remote control devices with the help of the light emitting diode (LED). The LED is used to emit infrared wave to the device that needs to be controlled by the remote control. Due to the limited range of infrared, it cannot send transmission signals through walls or any other obstacles. Infrared technology is also being used in personal data transmission devices, such as laptops, personal digital assistants (PDA), and mobile phones. A visual illustration of the uses of infrared in a remote device to control a smart television is shown in Diagram 11.2.

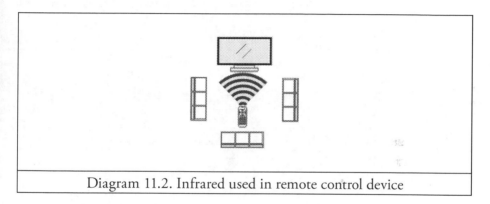

Diagram 11.2. Infrared used in remote control device

## 11.3   Radio Wave

Radio wave is the most common wireless channel used in today's modern era. It emits different ranges of frequency bands that are between 3 kHz to 300 GHz. Radio waves are essentially a part of electromagnetic radiation, a large group of waves that are broken down into separate smaller groups based on their wavelengths and frequencies. Radio waves are used in a variety of applications and they are the primary mechanism behind FM and AM radio broadcasting, television broadcasting, as well as paging and radar systems.

Radio waves have a limited frequency spectrum. Consequently, their operation is regulated by laws and any entity who wants to use radio waves for their operations need to obtain a license before they can legally continue with the use of this communication channel. While frequency is limited, though, radio waves can penetrate building walls and can transmit signals even when certain obstacles are present on their line of transmission. Radio waves often undergo refraction, reflection, scattering, and diffraction when transmitting in high traffic areas or in areas where obstacles are present. The additional drawback with this type of communication channel is that the quality of its transmission can be affected by atmospheric elements, such as rain, lightning, and radiation. These elements may lead to impairments or issues that can have significant effects on the quality of signal being sent to a receiving station. A visual representation of radio wave broadcasting from a tower is shown in Diagram 11.3.

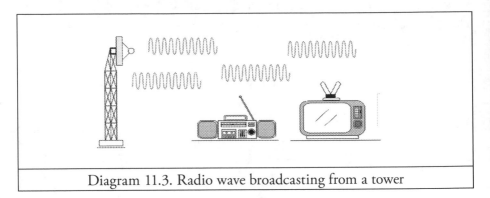

Diagram 11.3. Radio wave broadcasting from a tower

## 11.4   Microwave

Microwave is a type of electromagnetic wave that is relatively smaller than the waves used in radio broadcasting. Due to its short range, a microwave can even be measured in centimeters and millimeters. The wavelengths of a microwave can range from one millimeter to one meter, with equivalent frequencies between 0.3 GHz (300 MHz) and 300 GHz. Like the typical radio waves, microwaves can be broken down into multiple bands of frequency ranges for use in different applications. Microwaves can be used in both broadcast and non-broadcast procedures, such as point-to-point communications. Some of the most common applications of microwaves include satellite navigation, spacecraft communication, mobile communication, airborne radar, and international broadcasting.

Microwaves with high frequencies can speed up data transmission even with low bandwidth. However, channels using this wave may be susceptible to blocking interference, which could include atmospheric intrusions, such as solar radiation, snow, rain, and electromagnetic radiation. This problem, however, can be avoided by placing microwave antennas in high altitudes. An illustration of microwave broadcasting from an mobile tower to give coverage to certain radius of servicing area is shown in Diagram 11.4.

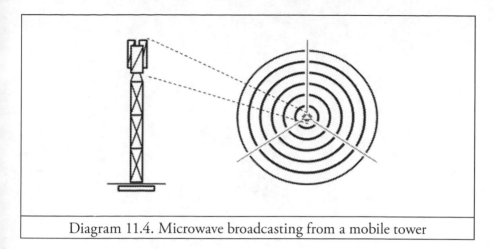

Diagram 11.4. Microwave broadcasting from a mobile tower

# Chapter 12

# Noise

## 12.1 Noise Overview

Noise is an unwanted disturbance in a communication channel. Its presence can obscure the transmission of any signal and prevent it from reaching its intended destination in a safe condition. Noise is categorized into different types in telephone communication. Types include thermal noise or white noise, cross talk noise, and impulse noise. These types of noises can significantly affect the overall quality of voice communications.

## 12.2  Thermal Noise/White Noise

Thermal noise is an electrical disturbance caused by the thermal agitation of electrons in conductors. All communication devices and transmission media are susceptible to this type of noise. Thermal noise is characterized by the difficulty in propagating signals due to the increase of kinetic energy followed by the random motion of electrons within the electrical devices. Thermal noise is also known as white noise due to its uniform distribution across the spectral frequency.

Thermal noise can be generated in transmission channels through terrestrial effects, including thunder attack, solar radiation, and aurora.

## 12.3 Crosstalk

Cross talk is a disturbance caused by electromagnetic interference (EMI), such as the undesired conductive, inductive, or capacitive coupling of transmission signals between channels or circuits. The presence of this type of noise in channels can hinder the transmission of data or cause errors in the packets being sent. An illustration of this phenomenon is shown in Diagram 12.3.

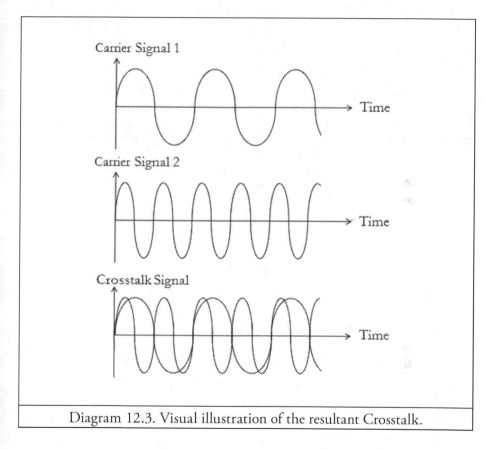

Diagram 12.3. Visual illustration of the resultant Crosstalk.

Crosstalk can occur in both wired and wireless channels. In a wired channel, multiple numbers of signal frequencies can travel in the same channel loop simultaneously and intersect with each other, resulting to impaired signal. In a wireless channel, crosstalk can occur when a signal from one wireless channel intersects with the signal coming from another wireless channel, resulting to intertwine data transmission. Conversations leaking into other people's connections in voice communications can occur in both wired and wireless channels that are experiencing crosstalk.

## 12.4  Impulse Noise

Impulse noise is a type of noise characterized by unwanted sharp sounds in the transmission channels. The sounds have irregular spikes caused by electromagnetic interference or by the agitation of electrons in electronic devices. Other causes include electromagnetic disturbance arising from solar radiation, aurora, lightning, and high tension in electrical cables. Impulse noise can occur instantly and can annoyingly increase the amplitude of carrier signals. A visual illustration of this type of noise in carrier signal is shown in Diagram 12.4.

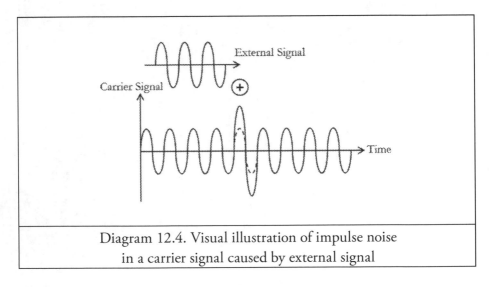

Diagram 12.4. Visual illustration of impulse noise
in a carrier signal caused by external signal

In analog signal transmission, impulse noise is considered a small issue, which can be removed or prevented with the help of a noise filter installed in the receiver's end of the network. In digital transmission, though, it is considered a high level problem as it can lead to the corruption of data stream.

# COMMUNICATION SYSTEMS

# Chapter 13

# Communication System

## 13.1 Fundamental of Communication System

A communication system is comprised of several elements that contribute to its overall function. An absence of even one of these elements can hinder the successful transfer of signals from one network to the other. A communication system commonly has five basic elements: the source, transmitter, channel, receiver, and destination. A block diagram of a basic communication system is shown below in Diagram 13.1.

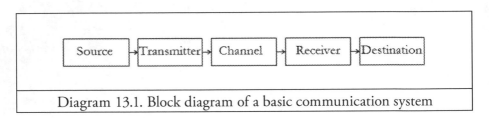

Diagram 13.1. Block diagram of a basic communication system

The elements in Diagram 13.1 are described as follows:

i.   Source -The origin of the data information.
ii.  Transmitter -The device that transmits the signal over the communication channel.
iii. Channel -A transmission medium that carries data information toward the receiver.
iv.  Receiver -A device used to receive signals from transmission channels.
v.   Destination -The endpoint or target location where the information being sent needs to go.

## 13.2 Analog Communication System

An analog communication system (ACS) is basically a set of system that is responsible for the orderly transmission of analog data signal from the sender to the receiver. A block diagram of this system is shown in Diagram 13.2.

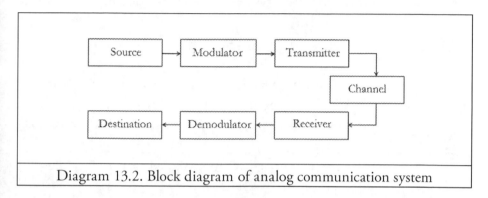

Diagram 13.2. Block diagram of analog communication system

An analog system contains the fundamental elements of a basic communication system, with two additional elements added for its intended function. These two additional elements include the modulator and the demodulator, which are explained below:

i.  Modulator -A mechanism used for modulating an input signal to carrier signal, so it can be sent suitably to the intended recipient via transmission channel. The modulator functions on the sender's end.
ii. Demodulator -A mechanism used to detect and demodulating carrier signal that received from the sending network to extract the original signal. The demodulator, on the other hand designed to work on the receiver's end.

An analog communication system has several advantages and disadvantages, which are explained below:

i.  Advantages

  a.  ACS requires only a small amount of bandwidth for the transmission of information.
  b.  ACS is less complex and can be easily managed.

ii.  Disadvantages

  a.  ACS does not allow the restoration of information that has been affected by noise at the receiver site.
  b.  ACS lacks security and protection, which means that private information sent over this system can easily be hacked.

## 13.3  Digital Communication System

A digital communication system is a set of system that can methodically transmit digital data signal from the sending network to the target network. A block diagram of how this system transmits information is shown in Diagram 13.3.

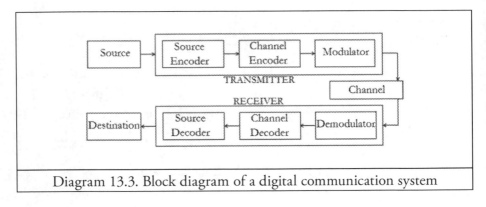

Diagram 13.3. Block diagram of a digital communication system

A digital communication system has the same elements found in an analog system. In addition it contains extra elements such as the source encoder, channel encoder, channel decoder, and source decoder. These elements are explained below:

i.  Source Encoder -A mechanism that removes all the redundancies in the data signal.
ii.  Channel Encoder -A mechanism that adds controlled redundancy to the data signal for later usage of erroneous data recovering process at the receiver end.
iii.  Channel Decoder -A mechanism that recovers the errors that occur in the data signal during the transmission process.
iv.  Source Decoder -A mechanism that removes the redundancy added by the channel encoder and restores the original data signal.

Like the analog communication system, the digital communication system (DCS) has its own advantages and disadvantage.

i.  Advantages

    a.  DCS allows the easy correction and restoration of the original transmission data.

    b.  DCS preserves privacy via data encryption method.

ii.  Disadvantage

    a.  DCS requires larger bandwidth for data transmission.

# Chapter 14

# Modulation

## 14.1 Modulation Overview

Modulation is a process by which the properties of a waveform or carrier signal are made varied for the successful transmission of information through a network. This modulation process can be divided into two types as analog modulation and digital modulation. Analog modulation is the process used for transferring center frequency of analog signal up to the high frequency of carrier signal. Digital modulation, on the other hand, is a method used for converting digital signal into analog signal during transmission. Below is a block diagram of the general modulation process.

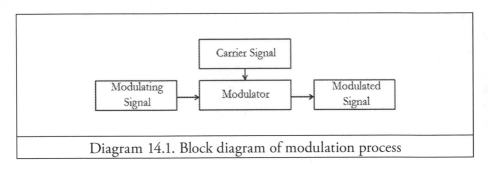

Diagram 14.1. Block diagram of modulation process

Both analog and digital modulation are categorized into different types. These are the following:

i.   Analog Modulation

    a.   Amplitude Modulation
    b.   Frequency Modulation
    c.   Phase Modulation

ii.  Digital Modulation

    a.  Amplitude Shift Keying
    b.  Frequency Shift Keying
    c.  Phase Shift Keying
    d.  Quadrate Phase Shift Keying

## 14.2  Amplitude Modulation

Amplitude modulation (AM) is a method used to move data onto a carrier waveform. One of the earliest modulation methodology used in radio communication, it works by varying the strength or amplitude of carriers based on the waveform being transmitted. While amplitude is made varied for the transmission, the carrier signal frequency itself remains at a constant level. The modulation process is visually illustrated in Diagram 14.2.

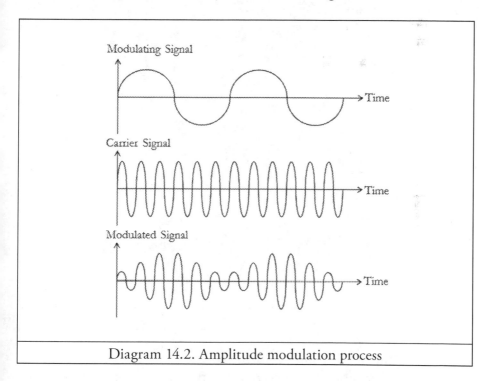

Diagram 14.2. Amplitude modulation process

Amplitude modulation uses only a small amount of bandwidth for successfully transmitting a broadcasting signal. The downside to this technique is that the signals it transmits are susceptible to electrical noises during the transmission process.

## 14.3  Frequency Modulation

Frequency modulation (FM) is a modulation technique used in communications wherein the frequency of a carrier signal is varied for signal transmission from the sender to the receiver. Unlike amplitude modulation where amplitude is varied for signal transmission, the amplitude of the carrier signal in frequency modulation remains constant and unchanged. Frequency modulation is visually represented in Diagram 14.3.

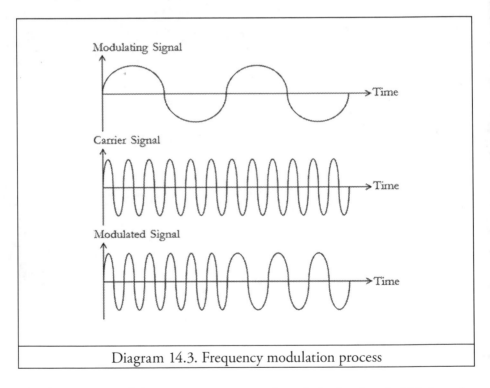

Diagram 14.3. Frequency modulation process

The frequency modulation process uses a relatively higher bandwidth in order to transmit broadcasting signal successfully. The advantage of this technique over amplitude modulation is that it maintains the high quality form of the original signal during the transmission process.

## 14.4   Phase Modulation

Phase modulation (PM) is a modulation technique that encrypts data signal as variations in the instantaneous phase of a modulated waveform. In this type of modulation technique, the amplitude of the carrier signal remains constant while the phase of the frequency varies. Unlike FM, phase modulation is not often used for radio wave transmission because it requires a more sophisticated hardware for receiving signals. The phase modulation process is visually illustrated in Diagram 14.4. In the diagram, it is shown that the modulating signal uses a carrier signal to produce a phase modulated signal.

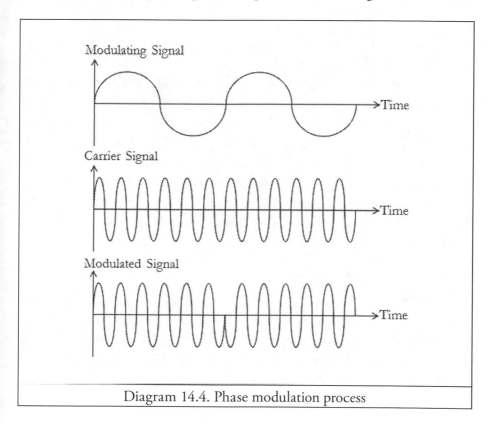

Diagram 14.4. Phase modulation process

# Chapter 15

# Multiplexing

## 15.1 Multiplexing Overview

In telecommunications, multiplexing is a process whereby multiple signals (analog or digital) are combined into one signal for transmission over a shared channel or transmission pathway. This technique is essential when transmitting multiple data over a shared transmission medium. Known for short as mux or muxing, multiplexing reduces the number of transmission lines required to transmit multiple data. It also maximizes the use of the bandwidth of a particular transmission channel. Multiplexing has provided several benefits to communication systems. One of the most common advantages it provides is that it makes data transmission more efficient in terms of cost, space, and materials used.

At times, the opposite of multiplexing may be required for a particular transmission process. Such process is called demultiplexing (also demux or demuxing). The demultiplexing procedure is used to separate the multiplexed signals at the receiver's network in order to decode information that has been sent. Both multiplexing and demultiplexing processes are visually illustrated in Diagram 15.1.

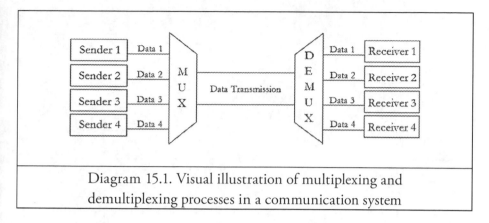

Diagram 15.1. Visual illustration of multiplexing and demultiplexing processes in a communication system

Different types of multiplexing are used in telecommunication networks. The use of each type depends on the requirements for sending data within a pathway. The different types of multiplexing include:

i.   Space division multiplexing access
ii.  Time division multiplexing access
iii. Frequency division multiplexing access
iv.  Wave division multiplexing access

## 15.2   Space Division Multiplexing Access

Space division multiplexing access (SDMA) is a method for accessing channels wherein spaces are created and assigned for multiplexing and demultiplexing data signals. The assignment of spaces is pre-defined in an orderly manner, resulting to data being transmitted with equal priority and without disturbance from any of the other existing data. A guard band is allocated between the spaces of each data signal so as to avoid interference and to make sure that data signal is transmitted without interloping with other present signals in the channel. A visual illustration of the SDMA process is shown in Diagram 15.2.

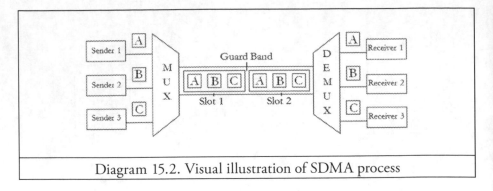

Diagram 15.2. Visual illustration of SDMA process

Data transmission using the SDMA method does not require dynamic coordination; however, it does require the precise alignment of networks to prevent errors during data transmission. Any error in data transmission can lead to wastage in resources and interference between or among channels, particularly for channels using the same frequencies.

## 15.3 Time Division Multiplexing Access

Time division multiplexing access (TDMA) is a channel access technique that permits several users to use the same frequency channel for information transmission. The transmission of various data signals by multiple users in a medium is made possible by the division of signals into different timeslots. Each user transmits data information within the time slot given to them. Transmissions are done one after the other in rapid succession to accommodate all users of the frequency channel.

The time division multiplexing access method is classified into two. These are the synchronous time division multiplexing access and asynchronous time division multiplexing.

i.  Synchronous Time Division Multiplexing Access

    a.  Diagram 15.3 (i) shows a visual illustration of the synchronous TDMA process.
    b.  In synchronous TDMA, the sender will be pre-allocated with a fixed time slot for data transmission.
    c.  Each time slot remains unoccupied if no data is being transferred.
    d.  If no sufficient data is available for transmission, bandwidth wastage may occur.

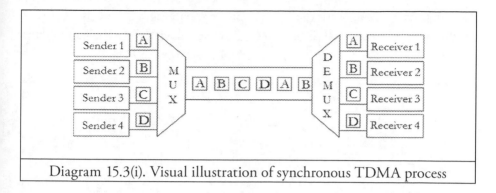

Diagram 15.3(i). Visual illustration of synchronous TDMA process

ii. Asynchronous Time Division Multiplexing

a. The asynchronous TDMA process is the complete opposite of the synchronous TDMA process.

b. Diagram 15.3(ii) shows a visual illustration of the asynchronous TDMA methodology.

c. In this system, the sender is allocated with a time slot for data transmission only when necessary.

d. All time slots is propagated with data only when there is a demand.

e. Bandwidth is used efficiently without wastage.

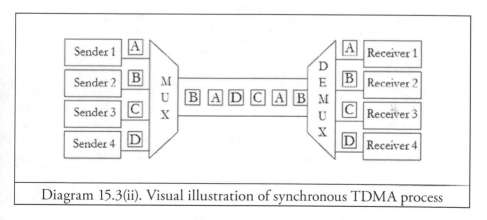

Diagram 15.3(ii). Visual illustration of synchronous TDMA process

## 15.4   Frequency Division Multiplexing Access

Frequency division multiplexing access (FDMA) is a channel access technique that combines several data signals of varying frequencies in a single channel and sends them simultaneously to their respective destinations. With this technology, users can share one transmission medium but are each given their own frequency for data transmission. With the FDMA process, data signals that use different bandwidth are transmitted through a single channel with the same bandwidth. A visual illustration of the FDMA process can be found in Diagram 15.4.

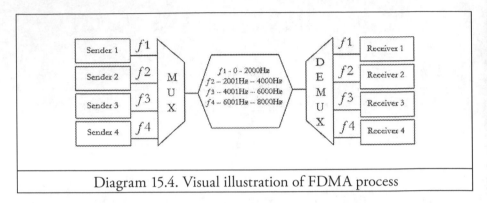

Diagram 15.4. Visual illustration of FDMA process

Data signals with varying frequencies that use the FDMA process for transmission are recovered through the demultiplexing method in the receiver's end. This makes it easier for the signals to reach their respective destinations successfully.

## 15.5   Wave Division Multiplexing Access

Wave division multiplexing access (WDMA) is a multiplexing technique that is widely used in fiber optics communication. This system is somehow similar to FDM in that it combines multiple signals for transmission over a single channel. However, instead of using radio frequencies (which is characteristic of FDMA) signals are sent through infrared (IR) transmission channel. Technically, the signals are combined on laser beams at varied IR wavelengths then sent along a fiber optic channel. A visual illustration of this process is shown in Diagram 15.5.

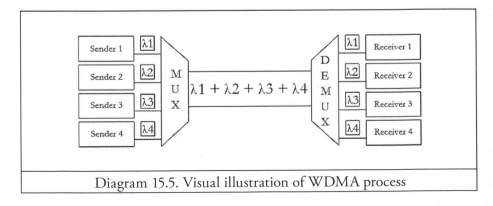

Diagram 15.5. Visual illustration of WDMA process

The wave division multiplexing access technique provides an efficient way to transfer data, particularly when it is used in conjunction with TDM or FDM. The combination of these methods allows data in different formats to be transmitted simultaneously in varying speeds through a single fiber optic cable. Theoretically, a strand of fiber optic cable can transmit data information rapidly at about several hundred gigabits per second (Gbps). At this speed, WDMA increases the efficiency of data transmission in long distance communication.

## 15.6   Code Division Multiplexing Access

Code division multiplexing access (CDMA) is a multiplexing method that allows users to send multiple data signals simultaneously over a single transmission medium. For this to be possible without any form of disturbance or interference among the various transmission data, spread spectrum technology is used by CDMA. A special coding scheme, wherein transmitters are each assigned a particular code for transmission, is also employed. A representation on how the CDMA process works is shown in Diagram 15.6.

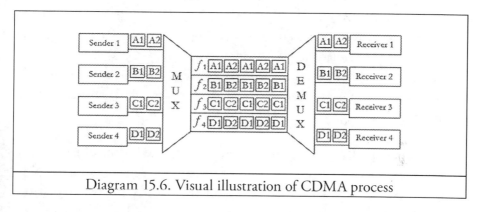

Diagram 15.6. Visual illustration of CDMA process

As seen in Diagram 15.6, four senders (Sender 1, 2, 3, and 4) are able to send signals simultaneously over a single medium and the signals are also received by the intended recipients at the same time.

Like any other multiplexing methods, CDMA enhances the use of a transmitter's available bandwidth. It has the ability to increase the capacity of data transmission by maintaining the signal to noise ratio. The CDMA technique is widely used in ultra-high-frequency (UHF) mobile telephone systems, as well as in third generation (3G) wireless mobile communications and second generation (2G) communications.

# Chapter 16

# Analog to Digital Conversion

## 16.1   Pulse Code Modulation

Pulse code modulation (PCM) is the process of digitizing analog signal into digital signal. This process is essential to transmit analog data in digital form via digital data network. For instance, in ISDN or mobile communication, the voice conversations will be converted to digital signal when transmitting over the digital network. Audio signals digitalized by means of PCM are known to be less susceptible to noise and interference, boosting their chances of getting received by their intended recipients without distortion or loss of information. Signals sent through PCM are also well protected, since the data sent can only be decoded using a special decoder designed particularly for the encoded signal being sent.

The transmission of digital signals is more sophisticated compared to analog transmission, but it provides a number of benefits that may not be found in the latter scheme. One known advantage of PCM is its ability to allow the storage of data in electronics devices. This technique is widely used in digital telephone recording systems as well as in other digital audio applications, including digital versatile discs (DVD) and compact discs (CD).

Analog signals undergo three different stages during their transformation into digital data. These include sampling, quantization, and encoding, as shown in Diagram 16.1.

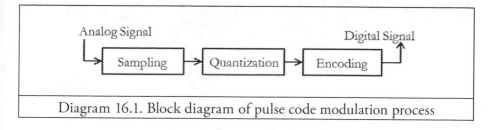

Diagram 16.1. Block diagram of pulse code modulation process

## 16.2 Sampling

Sampling is the first phase in the pulse code modulation process. At this stage, the amplitude of continuous time signals at discrete periods are measured and the continuous-time signals are transformed into discrete-time signals.

The sampling method has two types. One is natural sampling and the other is flat top sampling. In natural sampling the analog signals are reproduced with rectangular pulse trains and the analog segments are retained. The amplitude of the sampled pulse also varies depending on the amplitude of the analog waveform during transformation in the sampling period. In flat top sampling the sampled pulses have uniformed amplitude and the continuous analog signals are sliced into discrete values at regular time intervals, with the magnitude equal to the amplitude signals at a given period. The top of the sampled pulses take a flat shape in flat top sampling. Visual diagrams of both natural sampling and flat top sampling are shown in Diagrams 16.2(i) and 16.2(ii), respectively.

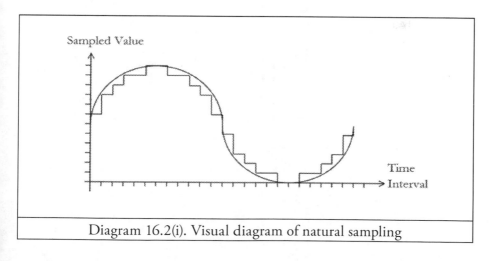

Diagram 16.2(i). Visual diagram of natural sampling

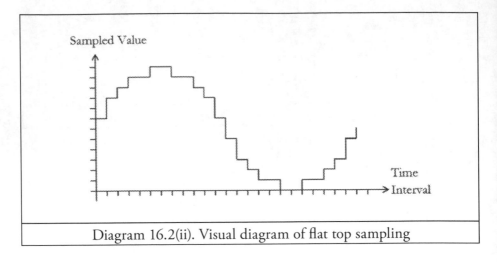

Diagram 16.2(ii). Visual diagram of flat top sampling

## 16.3  Quantization

Quantization is the second phase in the pulse code modulation process. In this stage, the sampled signals are aligned into equal time intervals. It is during this stage the continuous amplitude samples are converted into a discrete amplitude sample. The aligning of a sampled signal into balance time intervals is shown in Diagram 16.3.

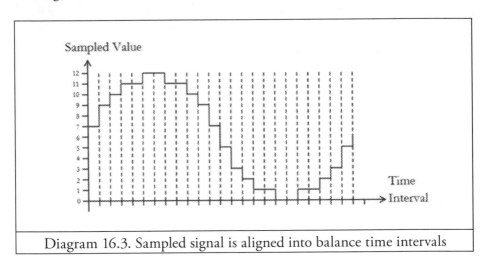

Diagram 16.3. Sampled signal is aligned into balance time intervals

# 16.4 Encoding

Encoding is the third phase in the pulse code modulation process. In this stage, the decimal values of the sampled signals are converted into binary values. The conversion of decimal values into binary values is shown in Table 16.3.

| Decimal Value | Binary Value | Decimal Value | Binary Value |
|---|---|---|---|
| 1 | 0001 | 7 | 0111 |
| 2 | 0010 | 8 | 1000 |
| 3 | 0011 | 9 | 1001 |
| 4 | 0100 | 10 | 1010 |
| 5 | 0101 | 11 | 1011 |
| 6 | 0110 | 12 | 1100 |
| Table 16.3 Conversion of Decimal value to binary value | | | |

Following the quantized signals shown in Table 16.3, the strings of the encoded binary values are 0111, 1001, 1010, 1011, 1011, 1100, 1100, 1011, 1011, 1010, 1001, 0111, 0101, 0011, 0010, 0001, 0001, 0000, 0001, 0001, 0010, 0011 and 0101 (7, 9, 10, 11, 11, 12, 12, 11, 11, 10, 9, 7, 5, 3, 2, 1, 1, 0, 0, 1, 1, 2, 3, 5), respectively.

# Chapter 17

# Digital to Analog Conversion

## 17.1 Overview of Digital to Analog Conversion

Years ago, the telephone communications industry used analog signal to convey information via networks. As technology advances most systems have switched to digital signaling as a primary form of communication. Nonetheless, analog still plays a crucial part in communication, despite the fact that new breakthroughs in technology have digitized most communication devices.

One instance that proves the importance of the analog signal in the modern era is that even though the voice information is sent as a digital signal; these signals are then converted back to analog for easier decoding of information. That is because the human ears can only decode voice signals transmitted in analog form.

Digital to analog conversion is commonplace in telephone communication systems. This done by using a process called digital modulation, which comes in different types. Types of digital modulation techniques include:

i.    Frequency Shift Keying
ii.   Amplitude Shift Keying
iii.  Phase Shift Keying
iv.   Quadrate Phase Shift Keying

These types of modulation techniques are used on applications that match their individual functions.

## 17.2   Frequency Shift Keying

Frequency shift keying (FSK) is a frequency modulation method used for the transmission of digital signals. In this system, two pairs of binary data (low and high, representing mark frequency 1 and space frequency 0, respectively) are transmitted using discrete frequencies. Frequency shift keying is done with the help of a modem, which converts binary data into FSK for transmission over a communication medium. During the process, the amplitude and phase of the signal remains constant but the frequency undergoes a change. The modem used for this process, which is known as the FSK modulator, can detect frequency shifts in carrier signals and determine their respective binary values efficiently.

At the receiving network's end, the binary values transmitted are decoded by a demodulator, which also recovers the original data information sent by the modulator from the sender's end. A visual diagram showing how FSK modulation works can be seen in Diagram 17.2.

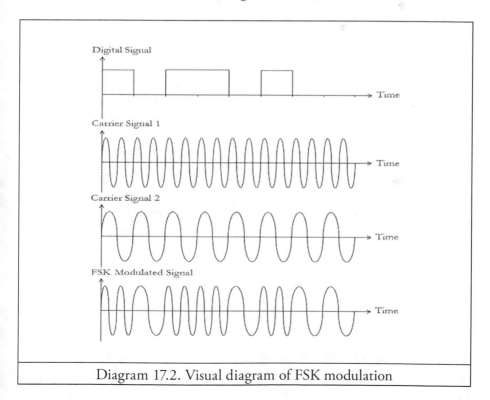

Diagram 17.2. Visual diagram of FSK modulation

There are several advantages to using FSK in transmitting digital signals. Two known advantages are its invulnerability to amplitude variations and its ability to prevent noise problems during transmission. It also does not require frequency accuracy to be absolute. Like any other modulation methods, though, FSK has its own downsides, one of which is its less bandwidth efficiency.

## 17.3   Amplitude Shift Keying

Amplitude shift keying (ASK) is the simplest form of amplitude modulation. In this technique, discrete amplitude signals are assigned bit values of 1 and 0, which represent high amplitude and low amplitude, respectively. The binary values are dictated by signal strength and throughout the process, the amplitude changes while the frequency and signal phase remain the same. A modulator and demodulator are often used to define the variations of amplitude in carrier signals based on the binary values when transmitting data information using this methodology. Diagram 17.3 provides a visual diagram of ASK modulation.

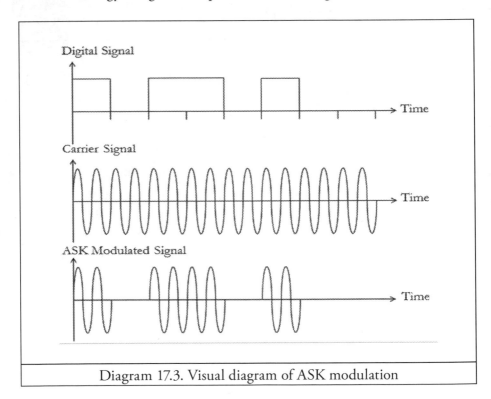

Diagram 17.3. Visual diagram of ASK modulation

The ASK process is often employed in fiber optic data transmission. Modulation and demodulation using this process is relatively inexpensive compared to its counterparts. However, some drawbacks are also known to be associated with this modulation method, the most common of which is its vulnerability to noise interference and its sensitivity to atmospheric and propagation conditions.

## 17.4   Phase Shift Keying

Phase shift keying (PSK) is a modulation process that shifts the period of a carrier wave for data transmission. The various instances of shifting a wave can represent bit values, such as 1 and 0, in signal transmission. Each wave shift means a change in the signal's phase, but in every variation of the phase, the amplitude and frequency remains constant and unchanged.

The simplest form of PSK is called binary phase shift keying or BPSK (also called phase reversal keying or 2PSK). In this method, two opposite signal phases 0° and 180° are used. The initial linear phase of the signal in this methodology is represented by 1 while the inverse phase of the signal is represented by 0. Apart from BPSK, there are also other forms of PSK, but they are more complex compared to the former.

The modulator and demodulator are designed with unique digital code patterns that can detect different signal phases. In the PSK method, the modulator encodes these patterns as binary symbols represented by particular phases. The demodulator on the other end of the line then recovers the original data sent by the modulator by determining the phase of the signal being received and mapping these phases to the symbols that they represent. A visual diagram of the PSK modulation is shown in Diagram 17.4.

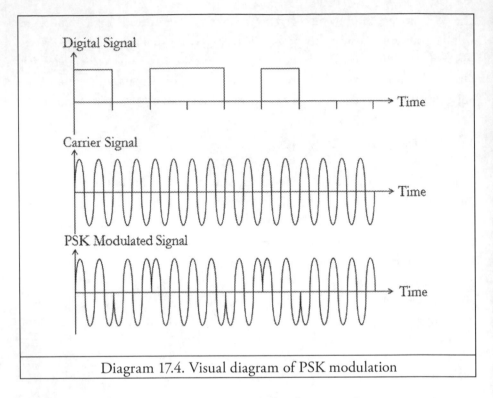

Diagram 17.4. Visual diagram of PSK modulation

The PSK process is widely used in many technologies, such as communication, due to its simplicity. It is also considered robust because of its non-susceptibility to noise and bandwidth limitation problems, which are characteristic of FSK and ASK.

## 17.5 Quadrate Phase Shift Keying

Quadrate phase shift keying (QPSK) is a modulation process that follows the same method of digital modulation employed by BPSK. In BPSK the shifting period of carrier waves are in in two phases only. But in QPSK the shifts periods of carrier waves are in four phases ($45°$, $135°$, $225°$, and $315°$). Moreover, QPSK utilizes a two-bit coding procedure instead of the single bit coding used by BPSK.

A visual illustration of the QPSK modulation process is presented in Diagram 17.5.

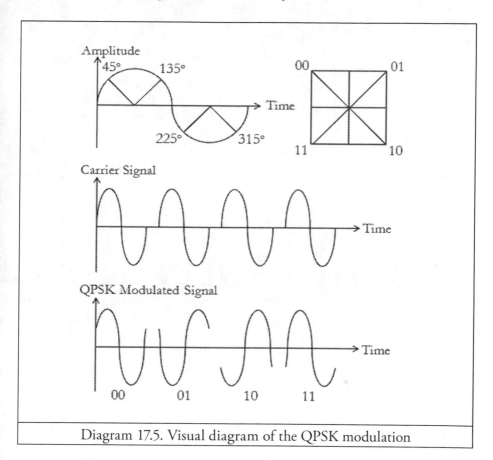

Diagram 17.5. Visual diagram of the QPSK modulation

# TELEPHONE NETWORKS

# Chapter 18

# Telephone Network

## 18.1   Plain Old Telephone Service

Plain Old Telephone Service (POTS) was the telephone communication system introduced in the beginning of the 20th century. It was among the telephone communication systems that were implemented using twisted pair wires. Widely utilized in many homes and small business offices around the world since its introduction to the public, this system runs at a low speed. POTS used circuit switching technology to provide bi-directional analog voice service.

POTS provided subscribers basic voice calling features, such as dial tones and calling tones. While it was based on analog signal transmission and had a relatively low bandwidth, and limited features compared to its more modern counterparts, it was known to be a reliable service. Consequently, it enjoyed popularity among many households and businesses.

Users of POTS were connected together via switching centers. Typically, subscribers were linked to the switching centers through pairs of overhead wires (called loops), which were interconnected to switchboards in the central office. The switching network that connected the subscribers to the central office was called customer loop.

Technically, this was how the POTS telephone system worked:

i.   Initially, Subscriber A would dial to telephone exchange office.
ii.  In the exchange office, a small light on a switch board would twinkle; indicating and informing the telephone operator that Subscriber A need service.
iii. Operator would manually connect to Subscriber A's customer loop, allowing the operator and Subscriber A to start speaking.
iv.  Subscriber A would tell the operator to initiate a call to Subscriber B by giving the phone number.
v.   Operator would look for the customer loop linking to Subscriber B, then operator would send a signal and ring Subscriber B's telephone.
vi.  Once Subscriber B answered the call, operator would establish a direct connection between Subscriber A and Subscriber B.
vii. At the end, the operator would gently disconnect himself/herself to ensure the privacy of the call sustained.

## 18.2  Public Switched Telephone Network (PSTN)

Public Switched Telephone Network (PSTN) is also known as POTS, but it is actually a more advanced form of the latter. While POTS offer only basic features, such as ringing signals and dial tones, PSTN has the added benefits of call waiting, speed dialing, and emergency call service.

It also offers an automated signaling and switching system. The circuit switching technology used by PSTN allows call coverage in an international level; that is, from the ability to connect subscribers within a local town, subscribers can now call love ones and friends from other parts of the globe. The PSTN system follows the standards set by the ITU-T for the seamless linking of telephone networks in different countries. It employs a strategic hierarchical numbering plan to handle thousands of calls from one location to various countries worldwide.

As the technology in telephone communication developed, PSTN gradually joined the digital bandwagon and begins to support packet switching network. This move allowed PSTN to provide additional values to voice, data, and fax services in same landline.

## 18.3  Integrated service Digital Network (ISDN)

Integrated Services Digital Network (ISDN) is a communication technology that marked the beginning of digital telephone networks. Primarily created as an alternative to the traditional dial up telephone networks in the 1990s, it allows the upgrade of traditional circuit switching facilities into packet switching systems, subsequently permitting the simultaneous digital transfer of data and voice signals over ordinary telephone twisted pair wires.

The key features of the ISDN system is its integrated circuits and processors, which help in the transmission of both voice and data signals. These elements are integrated into the switching machines found in the central offices of telephone networks. The switching machines used in the ISDN interfaces are made up of digital switches and they are designed to store subscriber information and call routing information in their electronic memory, allowing them to provide resourceful services to subscribers. Apart from permitting individuals to send text and voice messages, the robustness of the ISDN technology allows real-time video conferencing as well.

Two interface levels are used by the ISDN system. These are the Basic Rate Interface (BRI) and the Primary Rate Interface (PRI). The BRI is an entry-level interface and is targeted for use by small enterprises and homes. The PRI, on the other hand, is designed specifically for large users such as business organizations and offices. The BRI has a low transmission rate while the PRI provides a higher transmission rate.

Both the BRI and PRI interfaces are made up of B-channels and D-channels, which are responsible for carrying voice and data services, as well as signaling and control information, respectively. Specifically, the BRI has two bearer channels and one signaling channel (2B+D), which comprise of 64 Kbps each and 16 Kpbs respectively. The BRI has a transmission rate of 144 Kbps.

A visual model of the BRI and PRI interfaces shown in Diagram 18.3.

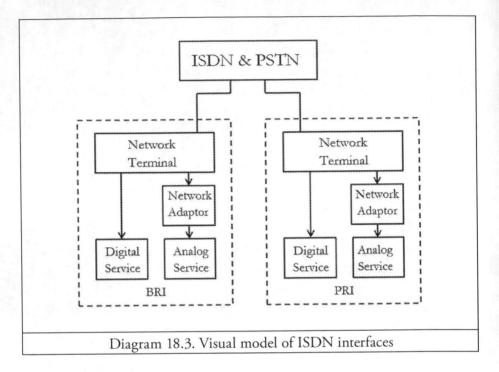

Diagram 18.3. Visual model of ISDN interfaces

On the other hand in PRI, the bearer channel and the signaling channel are comprised of 64 Kbps in each channel. The PRI has 30 bearer channels and two signaling channels (30B+2D) in Europe, and 23 bearer channels and one signaling channel (23B+D) in the US. Its transmission rate in Europe (30B+2D) is 2.048 Mbps, and in North America the transmission rate is at 1.544 Mbps (1.536 Mbps + 8bps).

Both the BRI and the PRI interfaces connect to ISDN by network terminals (NT). With the help of a network adaptor (NA), the connection allows the ISDN to provide direct digital services to subscribers or to offer analog transmission services.

The ISDN protocol provides an efficient call management system with low error rates and excellent communication security. Due to relatively high cost, ISDN struggle to penetrate the consumer market. The introduction of more affordable systems with higher speeds further contributed to its decline in demand.

# Chapter 19

# Landline Network Hierarchy

## 19.1 Landline Network Hierarchy

Landline telephony is a system comprised of direct physical lines that are individually connected to homes and offices. These physical lines are made up of metal wires, which serve as transmission channels. Landline telephones use dedicated lines. The routing, management, and maintenance of thousands of call connections between geographical networks with large jurisdictions are done in a strategic manner through a hierarchical system. A visual model of landline network with five hierarchy classes is shown in Diagram 19.1.

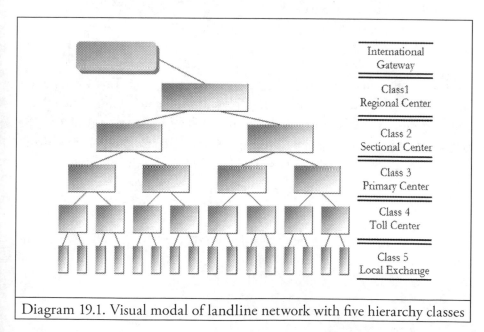

Diagram 19.1. Visual modal of landline network with five hierarchy classes

The roles of each class shown in the Diagram 19.1 are explained as follow:

*Class 1 Office -Regional Center*

The class 1 office in the hierarchy is the regional center. This class plays an important role in landline telephony for three reasons. First, it serves as the last resort for call routing when the lower centers in the hierarchy do not have direct links for call routing. Secondly, it is regulated by the communication authority and is run by staffs of engineers who are authorized in resolving higher order call routing and call management due to congestion, breakdown, and other emergencies. Thirdly, it serves as the collection point for landline cables that would be routed to the international gateways or special centers either via satellite or undersea fiber optics cable.

*Class 2 Office -Sectional Center*

The class 2 office in the hierarchy is the sectional center (SC), which is connected to the regional center and to all other sectional centers within one or two provinces of small states, or to a significant portion of a large province or state. In a nutshell, it makes possible inter-provincial or inter-state call connections. The SC is also responsible for connecting calls between primary centers that are below its hierarchy.

*Class 3 Office -Primary Center*

The class 3 office in the hierarchy is called as primary center. This central office handles calls that are made outside of a small geographical area where local exchange circuits are not directly linked to the class 4 office or toll center. The Primary Center is responsible for completing call connections between toll centers and is, therefore, the busiest among all offices in the hierarchy.

*Class 4 Office -Toll Center*

The class 4 office in the hierarchy is called as toll center. This office serve calls orderly connecting to other toll center through primary center or connecting to local exchanges in same area. Toll center is serving very important role in landline network that take account on calls connection charges based on the hierarchy passed through to complete each calls. Toll center also called as tandem office because all calls have to pass through this location to get to another part of network.

*Class 5 Office - Local Exchange or End Office*

The class 5 office in the hierarchy is called as local exchange office or end office. Among all other offices in the hierarchy, the local exchange has the closest connection to the customer's end. It is the office responsible for delivering dial tones to the customer.

## 19.2   Landline Numbering System

The landline numbering system was first created sometime in the 1940s. It debuted under the North American Numbering Plan. The logic behind the system was to provide a better way to identify subscribers and not have to go through a telephone operator when connecting calls from one network to another. Each subscriber was given an individual caller identification number, which can be used to directly contact them anytime. The landline numbering system not only made making calls easier and more efficient; it was also designed to systematically manage and maintain the unique caller identification numbers of all subscribers.

A landline telephone number is basically a set of uniquely sequenced digits. The very first telephone numbers used were relatively short, containing at least one, two, or three digits. As years passed, these numbers became longer, but they still serve the same purpose. Technically, landline numbers are pre-regulated to contain information that would identify the intended destination of a telephone call. A typical landline telephone number will have information that can identify the particular country, state, town, district, and local loop to which the call is being made.

For instance, an individual from the United States may want to place a call to a person located somewhere in London, England. The recipient has a number with this sequence: 011 44 491 577 031. The breakdown of this number would be:

i.   011: indicates an international call to the telephone network system
ii.  44: country code for England
iii. 491: city code for London
iv.  577: local exchange within London
v.   031: particular subscriber phone number.

The telephone network of the calling country will recognize 011 as an international call and will route the call to England, since the next digits are 44, which is England's country code. The rest of the numbers will determine and transfer the call to the telephone network of the recipient to complete the call.

# Chapter 20

# Private Telephone Exchanges

## 20.1  PBX

PBX stands for private branch exchange, a telephone switching system used by private enterprises or organizations. Unlike the traditional forms of telephone exchange, which are operated and owned by a telephone company, this mechanism is owned and operated by the organizations themselves. Originally, this machinery used the analog system, but in these modern times, it now uses digital technology.

A PBX can handle multiple call processes simultaneously. It can make the calls, receive calls, handle conference calls, and serve internal call routing. It also offers call waiting and speed dialing services for internal contacts as well as to numbers outside of the internal network. The PBX central office lines connect to a public switched telephone network by means of fiber optic channels or through copper wires. The system used by PBX allows several numbers of stations to establish connections simultaneously and to share phone line. It also allows two or more stations to engage in conferencing calls without the use of a central office equipment. This feature is made possible through the use of a special hardware or software control program, which is built within the PBX unit of the subscriber.

Typically, a PBX unit receives few phone lines from service provider and split the phone lines into multiple lines, that to be assigned to multiple numbers of users within an organization. The assignment of phone line is done in an orderly structure and each user is given a call ID. For clear understanding, let say an organization occupying with different department in each floors within a building, the each user in department will be assigned with own call IDs and allows them to connect with their colleagues from other departments located in other floors using the PBX.

The ability to establish a connection is not limited to the people within the building alone. It also permits the establishment of calls outside each department and outside the building used by the organization. This is made possible by using the PBX as a middle network that will direct the calls to public telephone networks outside of the system used in the organization. A computerize system will manage the switching of calls both sent from it and out of it and within it. Sometimes, it may also come with an optional feature-a switchboard or console designed for a human operator.

This system permits companies and businesses to handle several hundreds of calls simultaneously, regardless of whether those calls are made within or outside of the organization. Consequently, businesses are able to save on costs since they no longer have to get hundreds of phone lines from the service provider to be able to meet their communication needs.

In some instances, businesses and organizations may opt for an alternative to the PBX system. These alternatives include the key telephone systems, and Centrex service.

## 20.2  Centrex

Centrex is acronym for central office exchange service. It is a centralized telephone exchange system that works just like the PBX with a few distinctions. Like the PBX, it is designed for use by big business companies, organizations and campuses that require several thousand of phone lines. Centrex is owned, controlled and managed by the telephone line service provider. Basically, the equipment used in this type of telephone exchange is housed near to the location of the subscriber.

Centrex offers voice calls, call routing, call waiting, speed dialing, and conference calls. It also offers cost-accounting, call monitoring, direct inward dialing (DID), system sharing among multiple locations, and self-managed line allocations. The ability of Centrex to be managed by the service provider without putting the responsibility of maintaining the telephone exchange by the subscriber has made it as good alternative for PBX for larger organizations. Visual illustrations of both PBX and Centrex services are shown in Diagram 20.2(i) and 20.2(ii), respectively.

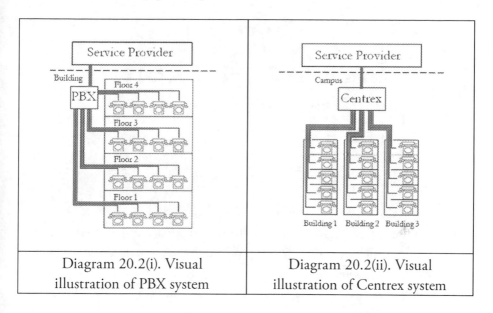

| Diagram 20.2(i). Visual illustration of PBX system | Diagram 20.2(ii). Visual illustration of Centrex system |

# Chapter 21

# Quality of Service (QOS) in Telephone Network

## 21.1   Telephone Traffic

Telephone traffic refers to the number of calls pass through a telephone exchange office or telephone network within a particular time interval. It can be likened to vehicular traffic in city roads and can vary significantly by the hour, days, weeks, months or years. Telephone traffic is one of the many aspects that telephone service providers take into consideration before establishing a telephone network or upgrading subscribers' telephone lines. The traffic of a telephone network can potentially affect a provider's reliability and quality of service.

Studying and monitoring the telephone traffic of their networks is a common undertaking performed by telephone providers. Knowing the traffic of their own telephone networks, assists telephone service providers to determine the state of their services.

In overall, telephone traffic is important due to the following:

i.   Telephone traffic helps in determining the condition under which adequate services are provided to subscribers while making economical use of the resources.
ii.  Telephone traffic is a basic analysis tool used for determining the requirement of providing a particular level of service for a given traffic pattern and volume.

iii. Telephone call success rate is important as it helps to determine the ability of a telecomm network to carry a given traffic at a particular loss probability.

iv. Telephone traffic serves as a key factor in estimating the cost effectiveness of a telephone company's infrastructure development model, allowing them to provide subscribers with the best accessibility for greater utilization.

## 21.2  Quality of Service in a Telephone Network

Quality of service (QoS) refers to the overall performance of a telephone service network as perceived by its subscribers. This aspect in telephone exchange is an important factor for telephone service providers to determine whether they have made progress with their business or not. Fundamentally, the QoS of a telephone service is measured by various elements, including service stability, service availability, reliability, and error rates to name a few.

Several factors can influence the QoS of a particular telephone service. These include:

i.   Traffic Pattern

    a.   Busy Hour- Traffic volume in a maximum stage for 60 minutes of time interval in a telephone network.

    b.   Day to day hour traffic ratio- Ratio of busy hour calling rate to the average day calling rate.

    c.   Seasonal traffic- Variation in the call patterns within a particular seasonal time scale.

ii.  Traffic Statistic

    a.   Calling rate- Average number of call requests from subscribers to make connections in per unit time.

    b.   Holding time- Average duration of the occupancy of a traffic path by a subscriber.

    c.   Occupancy- Average flow of call traffic in a communication server.

    d.   Traffic intensity- The volume of traffic passing on a transmission medium per unit time.

 e. Call completion rate- Ratio of successful calls to the number of call attempts.

 f. Congestion- Blocking of signals, hindering calls from passing through the transmission network. This can be due to the high number of call requests within the same time interval.

 g. Grade of service (GOS)- The proportion of unsuccessful calls relative to the total number of calls.